Electricity

Linda Howe

Contents

In the kitchen

Look at all the machines we can use in the kitchen. The machines look different, and they do different jobs. But in one way they are all the same. They all need electricity to make them work.

What is electricity?

Electricity is a kind of energy. Although you can't see electricity, you can see what it does.

What does electricity do?

Electricity makes things work. It warms your home and makes the toast. It turns on the light, the computer and the CD player. It even makes the phone ring.

Life with electricity

Electricity helps us in so many ways that it's hard to imagine life without it.

How does electricity help us?
Electricity gives us light and heat. When we walk into a cold dark room, we can turn on the light and the fire with the flick of a switch. Also, many machines run on electricity. They help us do our work quickly and easily.

Where is electricity used?
Electricity is used nearly everywhere – at home and school, in shops and offices, in hospitals, factories and on farms.

新宿
新宿 サンパーク
NEW SUNPARK
クスリ
龍生
MITSUMINE
新宿 サンパーク 本館
クスリの龍生堂薬局
カメラ 太陽堂
Konica
COSMO
50
東京信用

Electric lighting

We use electric lighting both inside and outside. Lights help us see when it's dark, and they make our homes more comfortable.

What lights do we use inside?

We use different lights for different jobs. Bright lights in the kitchen help us work safely. Spot lights and desk lights give us light where we need it. Table lamps have a softer glow. They make the room cosy.

What lights do we use outside?

When it's dark outside, many roads have bright street lights and road signs. Cars and bikes have lights, too.

Electric heating

We use electricity to heat our homes, our food and our water.

Heating a home

Electricity is just one fuel that we use to heat our homes. Other fuels are oil, gas, coal and wood.

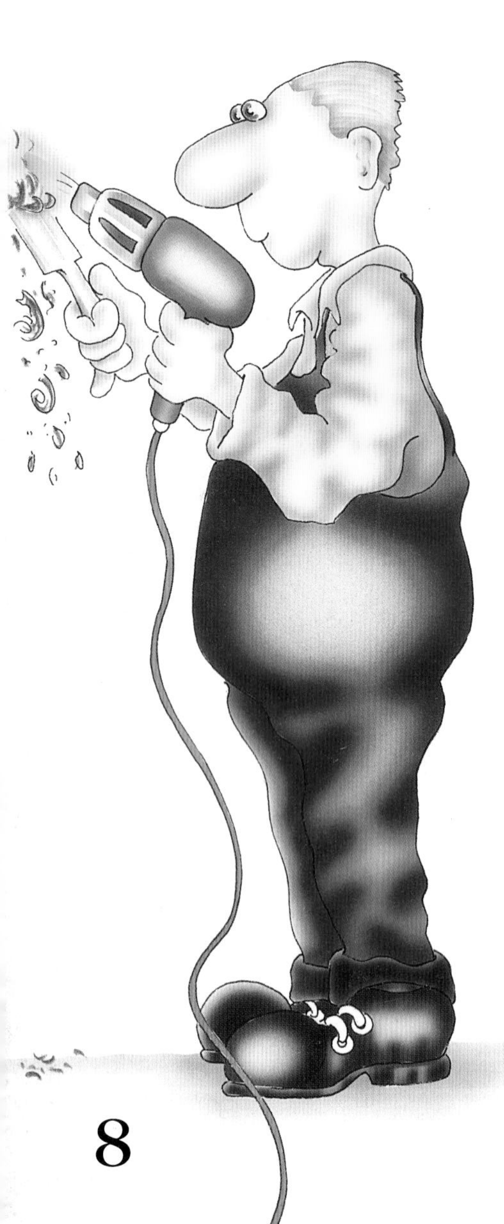

Heating food and water

We can use an electric cooker to cook our food. The oven and the hotplates are heated by electricity. Microwave ovens use electricity to heat food, too.

Electricity can also heat our water. We boil water in a kettle. Our bath water is heated in a tank.

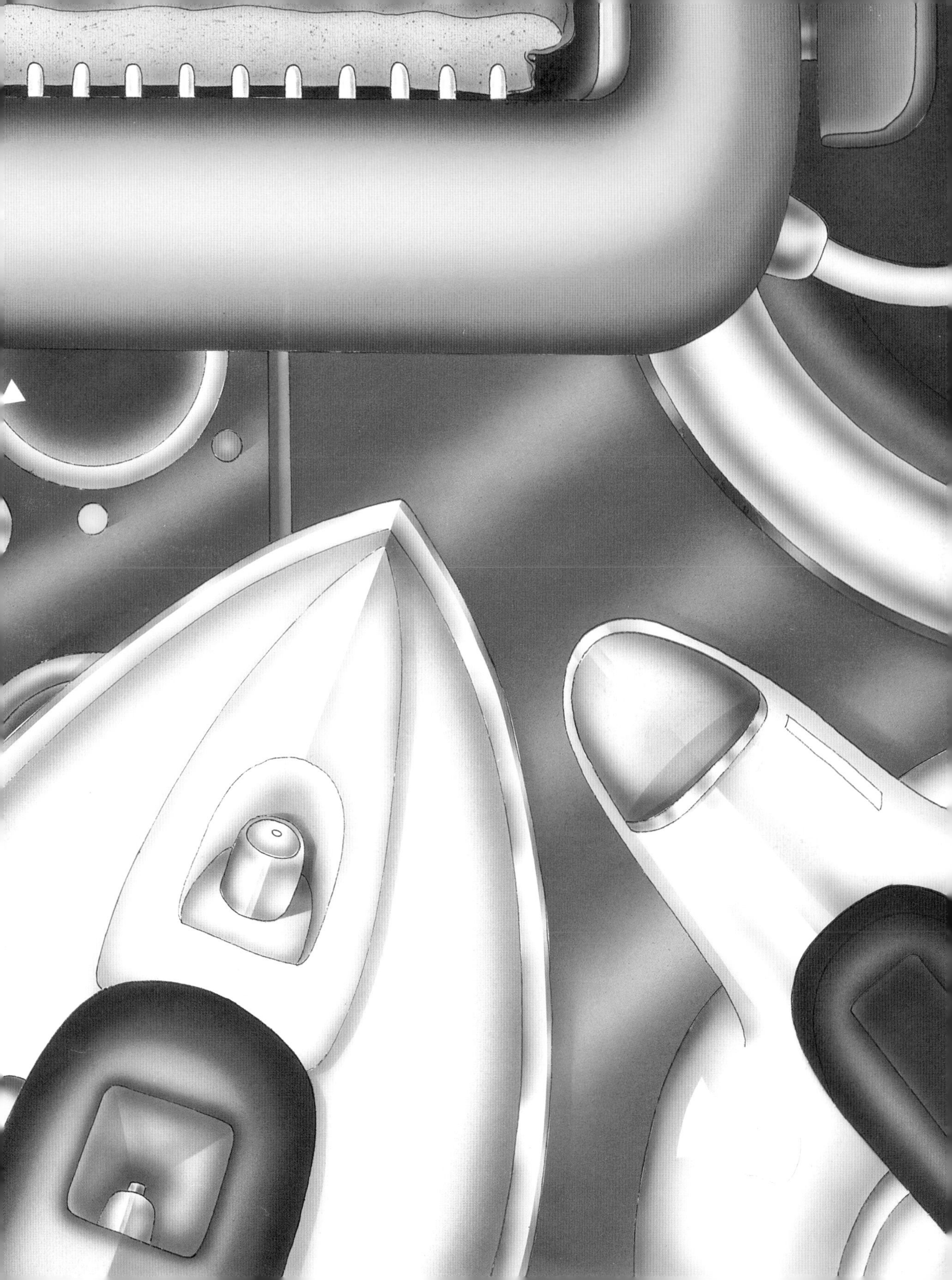

Electric machines

Electric machines help us do our work quickly. Many of them have moving parts that are driven by an electric motor.

Which machines have motors?
A washing machine has an electric motor that turns the drum. It will wash a pile of clothes in less than an hour.

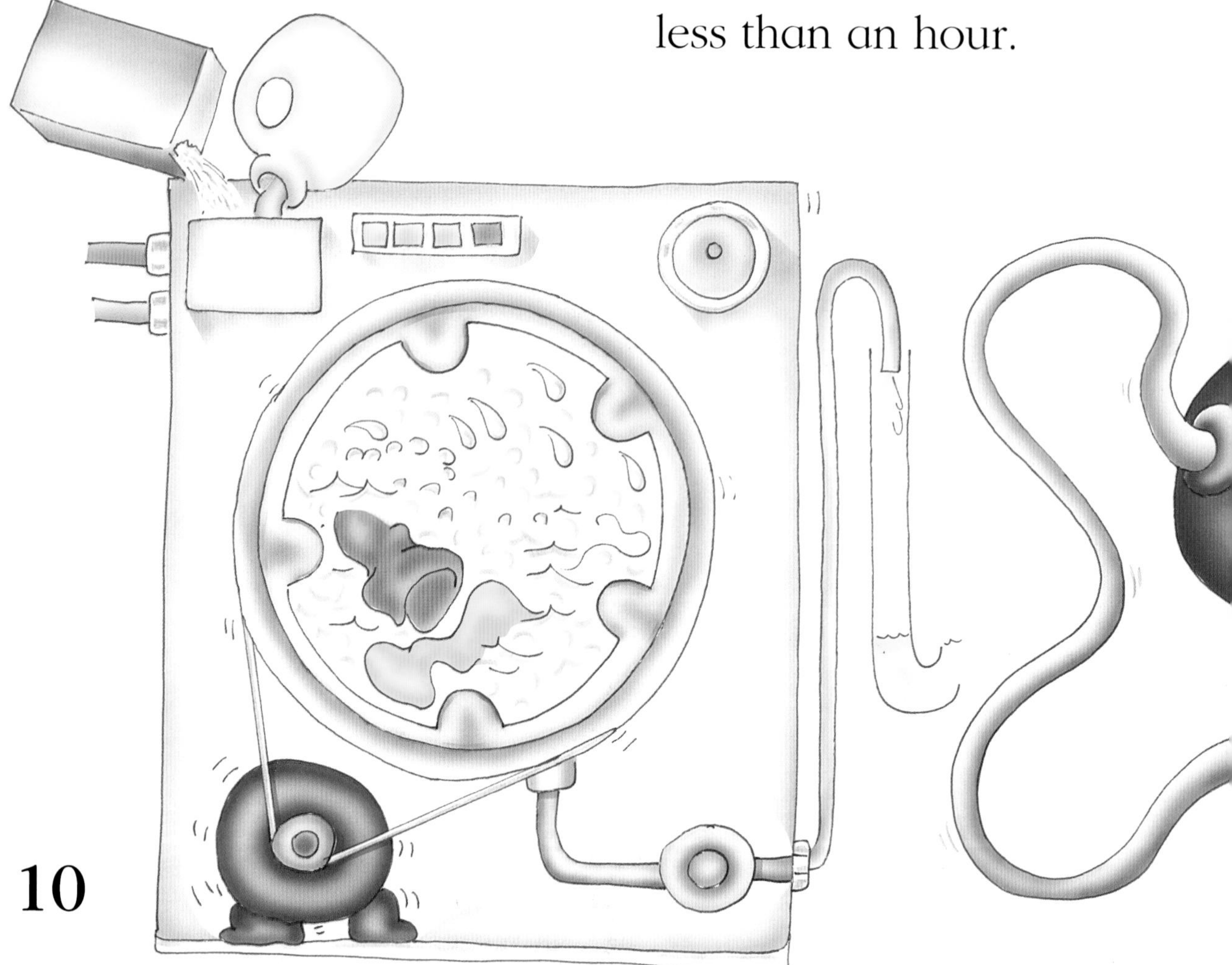

A vacuum cleaner has an electric motor, too. It spins a fan which sucks air into the cleaner – along with all the dust and dirt.

A dishwasher heats the water, pumps it round and blows hot air – all using electricity.

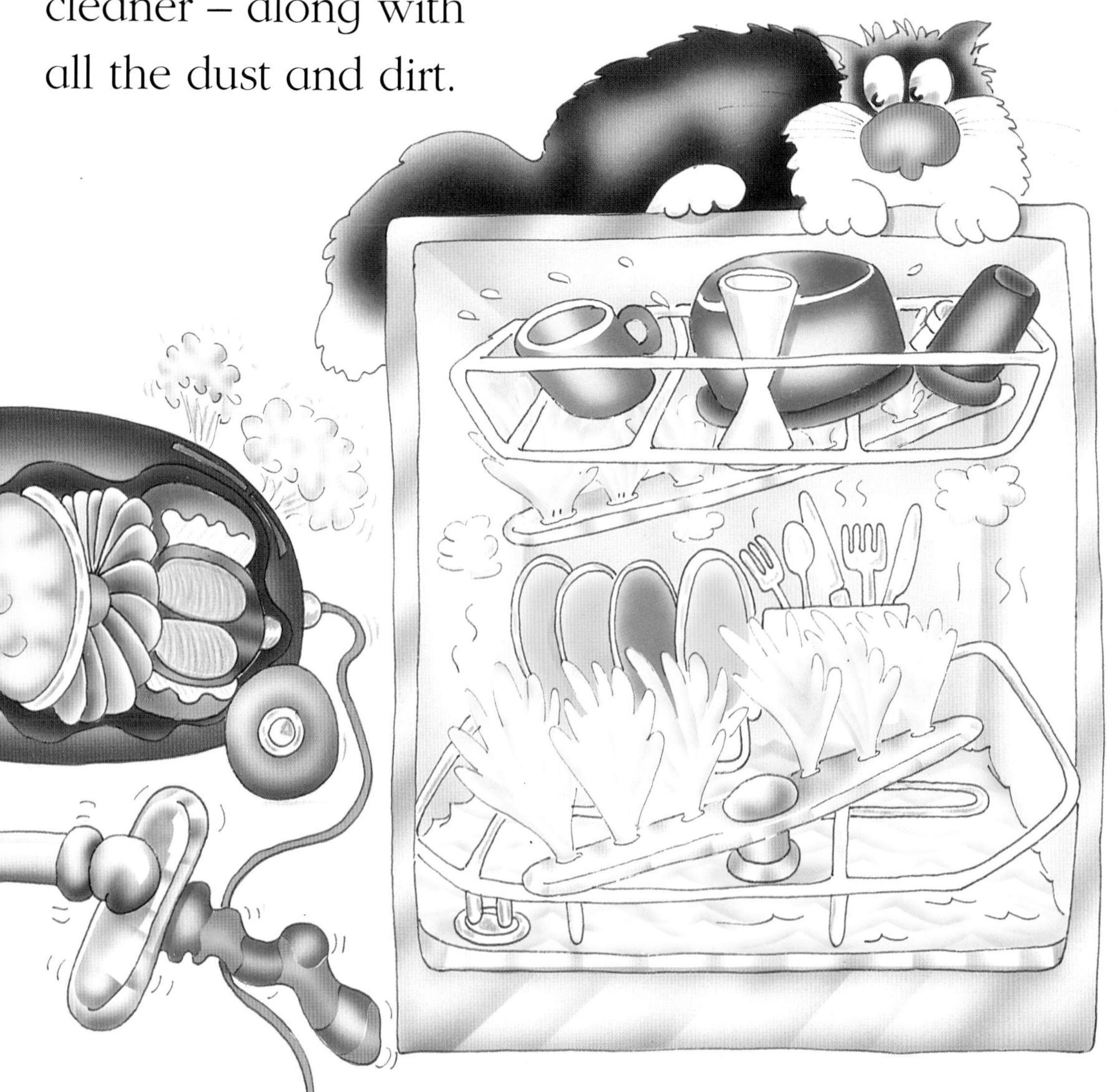

Batteries

What are batteries?

Batteries are like small parcels of electricity that can be moved from place to place. They are useful for toys, tools and torches. A good thing about batteries is that most are very safe.

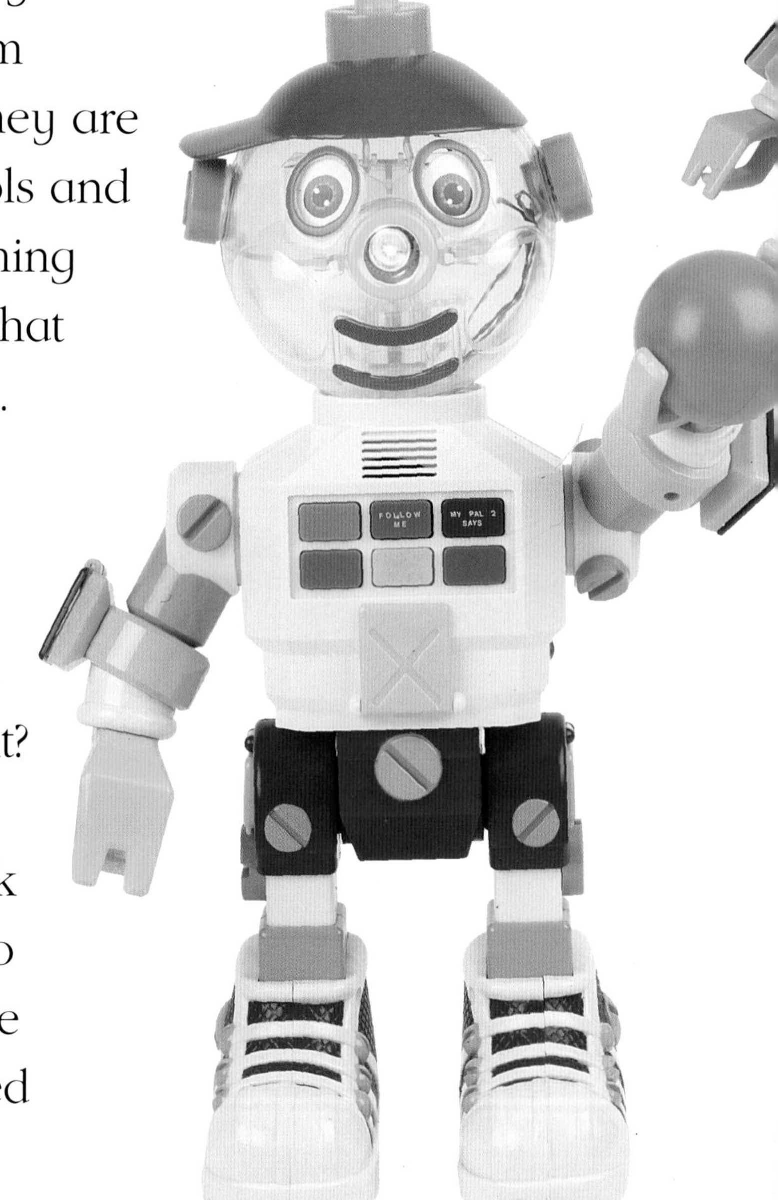

Do batteries last forever?

Have you ever left a torch on all night? By morning the light does not work because there is no electricity left in the batteries. You need to buy new ones.

Some batteries can be recharged. Putting them into a recharger gives them more power.

Before electricity

We have not always had electricity in our homes. Before, people had to use different kinds of energy to give them light and warmth.

What did people use before electricity?
Have you ever been in a power cut? Before electricity, life was a little like that.

People had to use candles, oil lamps and gas lights to light their homes. The lights weren't very bright and could easily start a fire.

People cooked and kept warm by burning coal or wood in stoves or on open fires.

Making electricity

Electricity is made in power stations. Some power stations use coal, oil, gas or nuclear power to make electricity. Others use moving water or the wind.

How does electricity reach our homes?

When electricity leaves a power station, it is carried along thick wires called cables. They are buried underground or hang on tall towers called pylons. The cables stretch for hundreds of kilometres, carrying electricity to towns and cities. In the towns, the electricity is spread out along smaller cables and flows safely into our homes.

Electric pathways

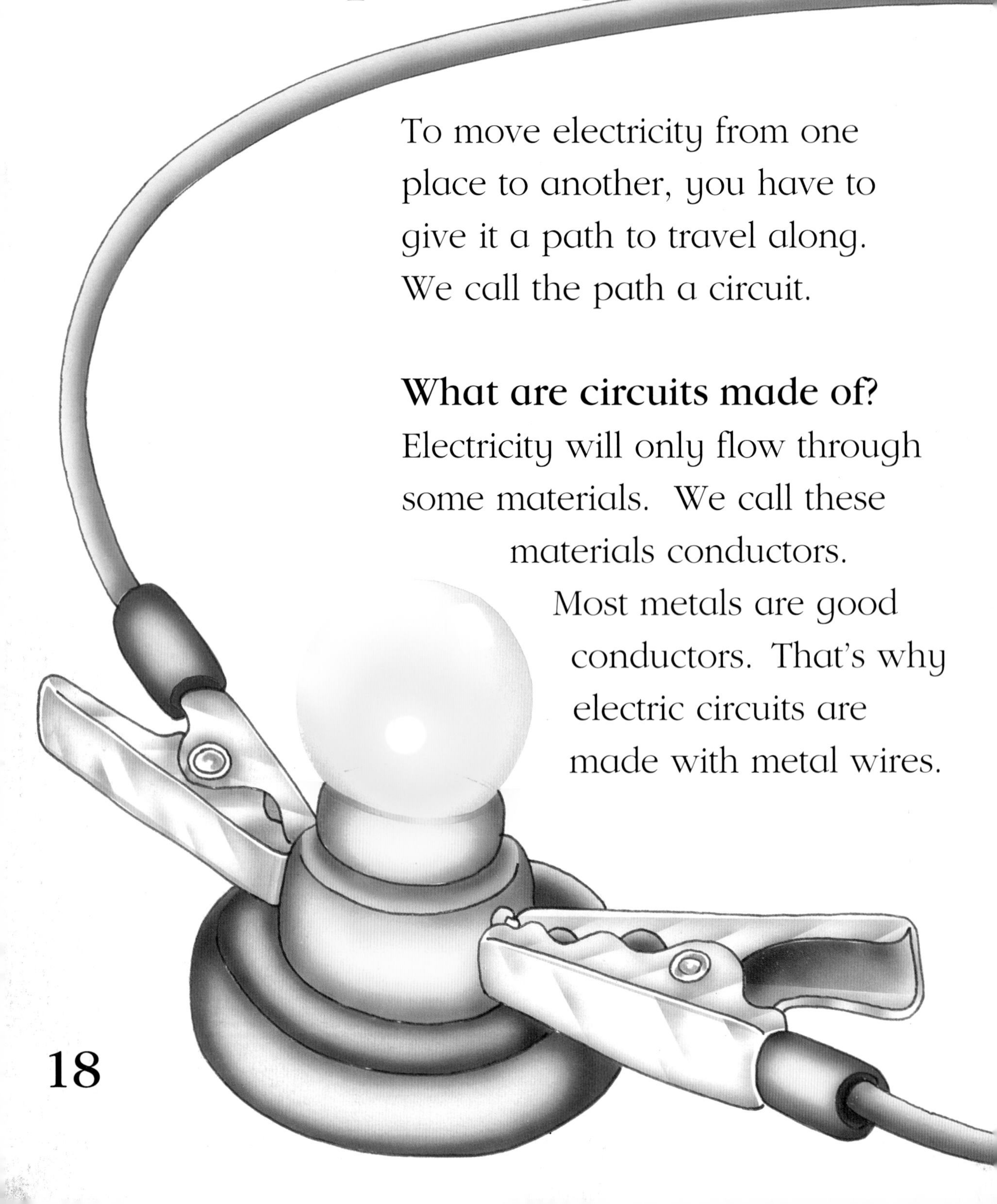

To move electricity from one place to another, you have to give it a path to travel along. We call the path a circuit.

What are circuits made of?
Electricity will only flow through some materials. We call these materials conductors. Most metals are good conductors. That's why electric circuits are made with metal wires.

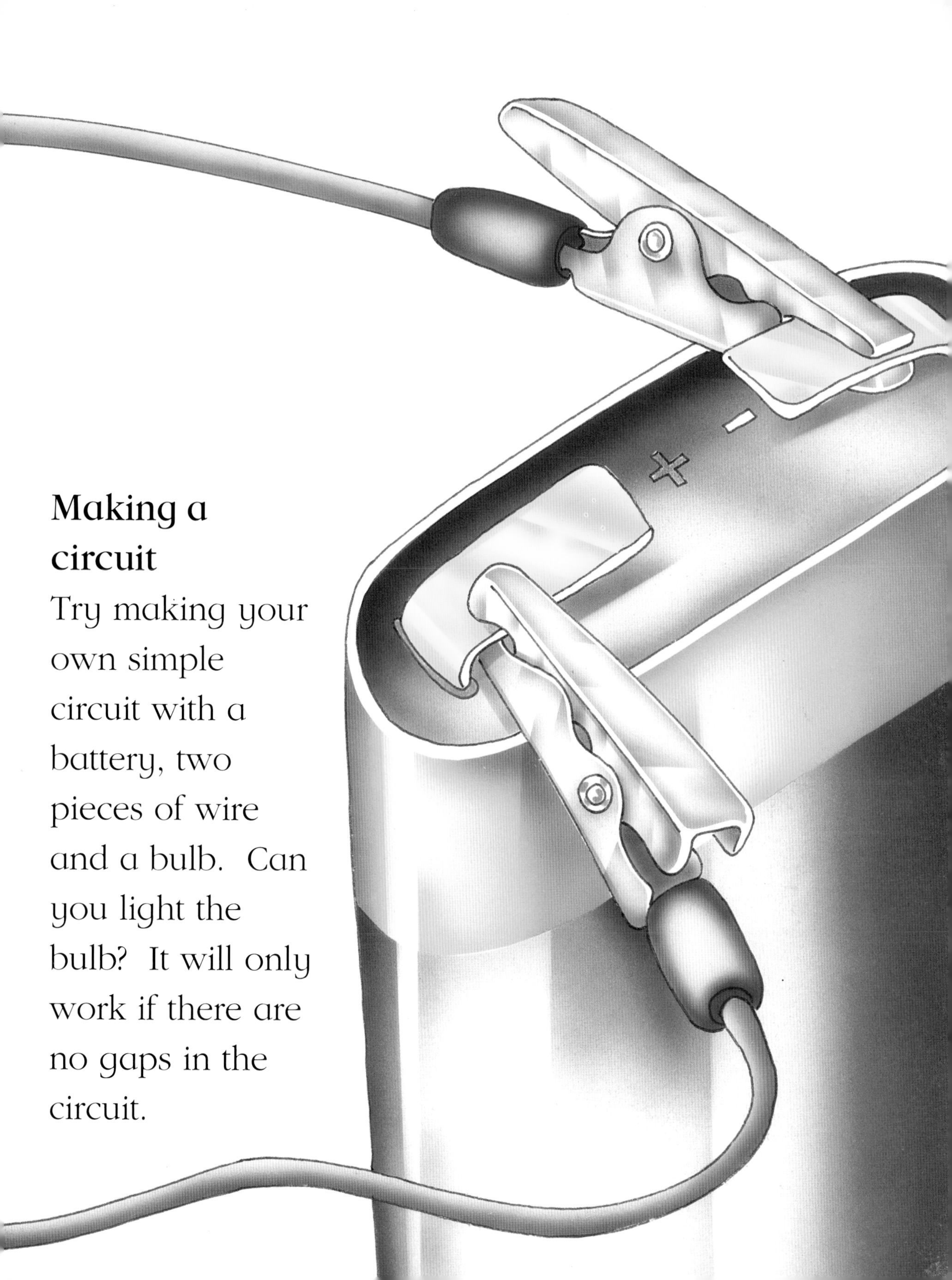

Making a circuit

Try making your own simple circuit with a battery, two pieces of wire and a bulb. Can you light the bulb? It will only work if there are no gaps in the circuit.

Stopping the flow

Electricity cannot flow through every kind of material. It cannot flow through plastic, wood or rubber. These kinds of materials are called insulators.

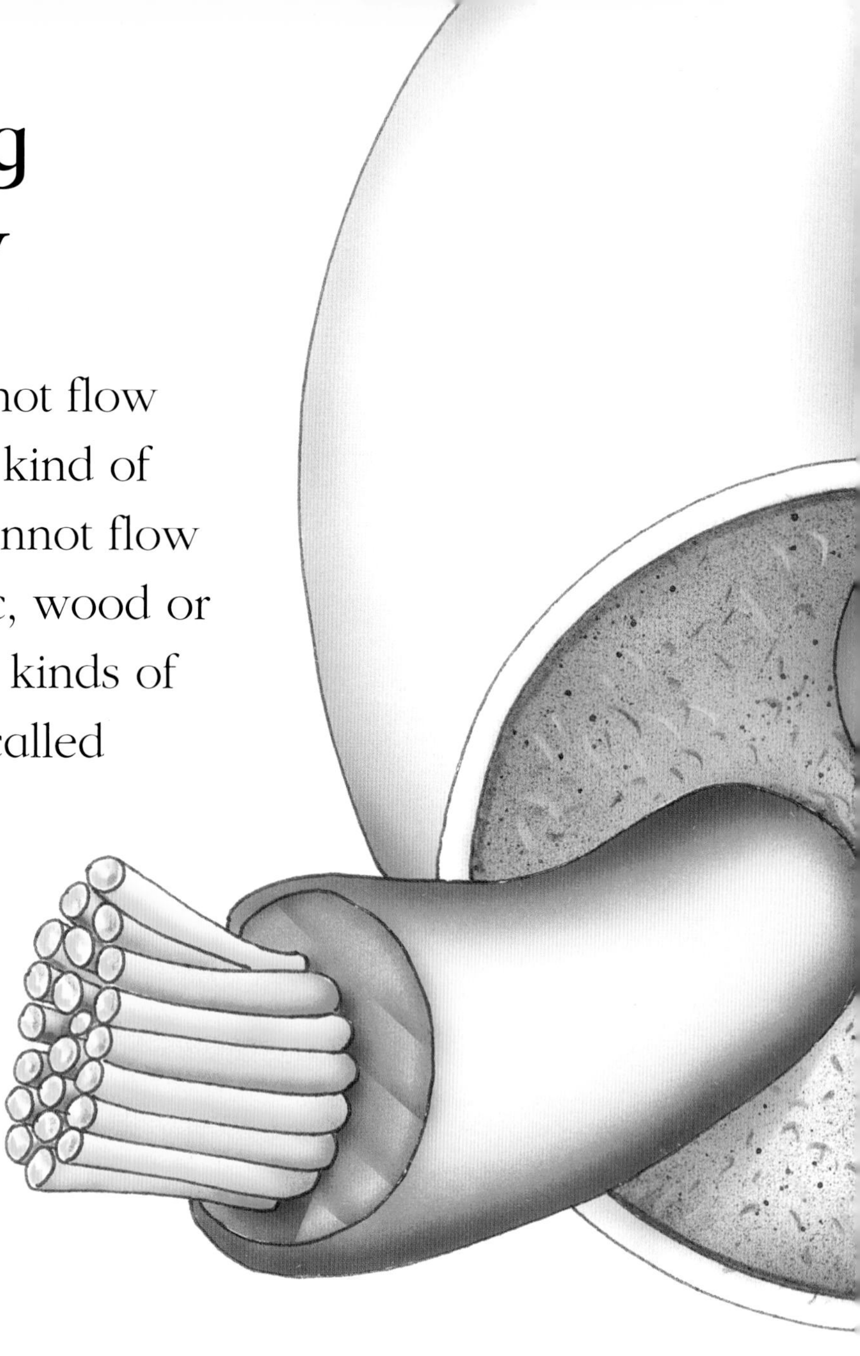

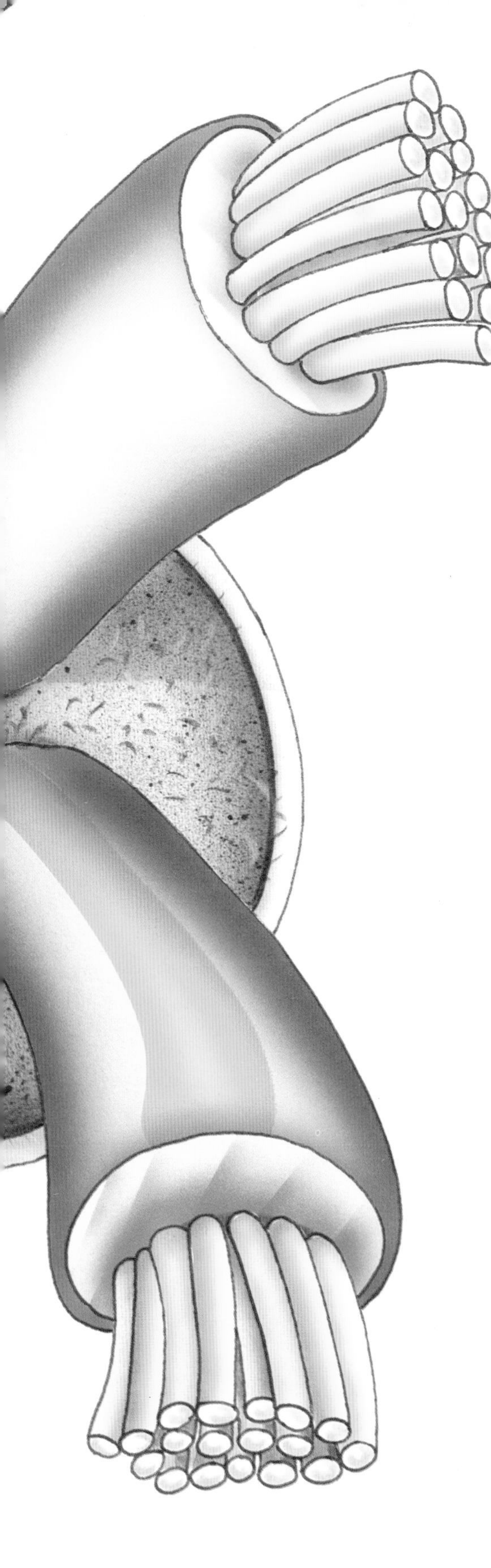

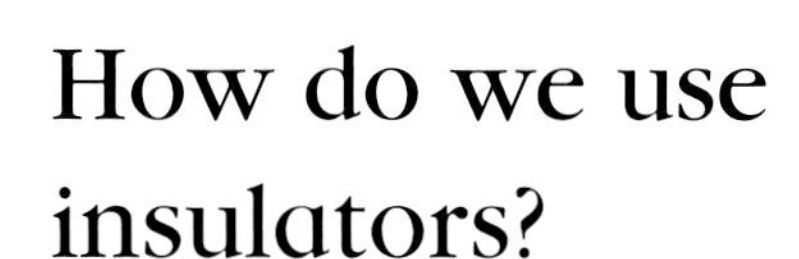

How do we use insulators?

Insulators help us to use electricity safely. They stop electricity leaking out of a circuit and hurting us.

Looking at flex

Inside a piece of flex there are metal wires. In a circuit, electricity flows along the wires to where it is wanted. The plastic tube around the wires is an insulator. It stops the electricity escaping.

Electricity at home

All round your home there are wires. They are hidden under the floors and behind the walls.

What are the wires for?
The wires make circuits for electricity to travel along. Some circuits carry electricity to the lights. Others carry it to the sockets.

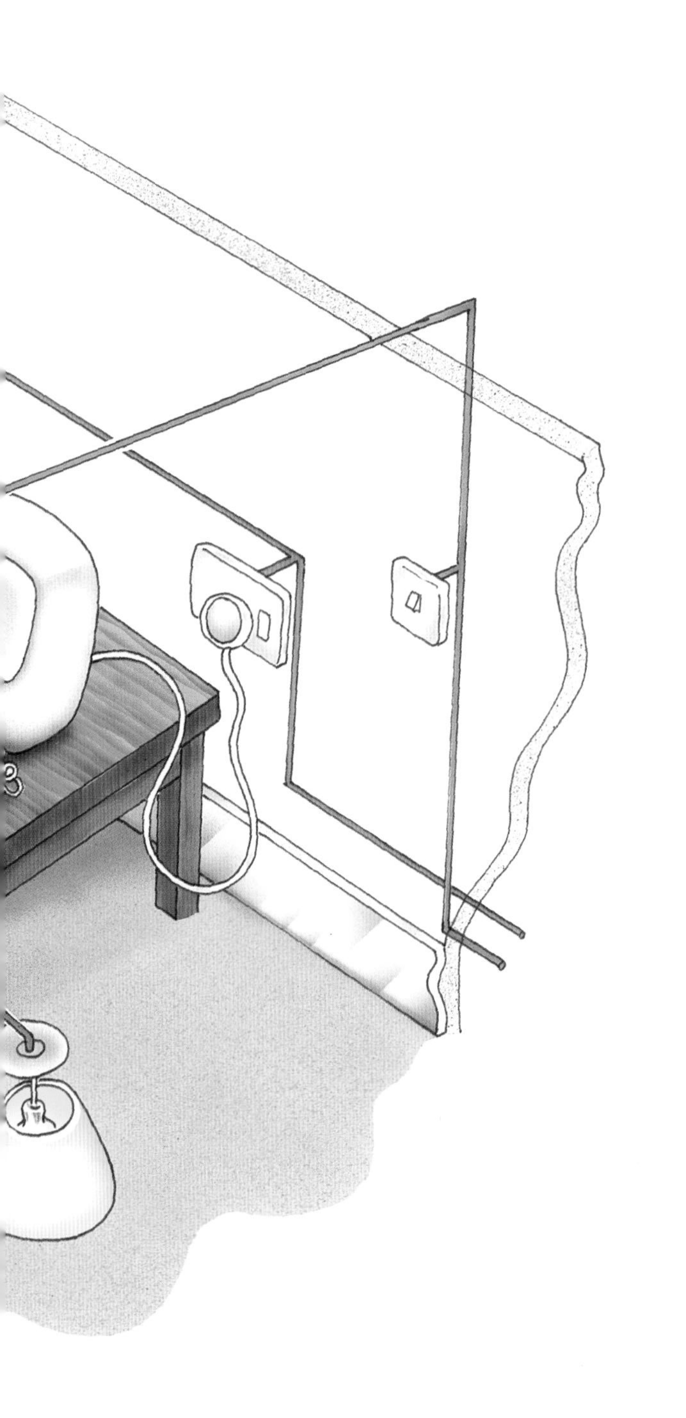

How do sockets work?

A socket is a meeting place. It's where an electric machine, like a television, can join the hidden circuit behind the wall. All the machine needs is a plug. When the plug is pushed into the socket, electricity can flow into the machine.

On and off

We turn machines on and off with the flick of a switch.

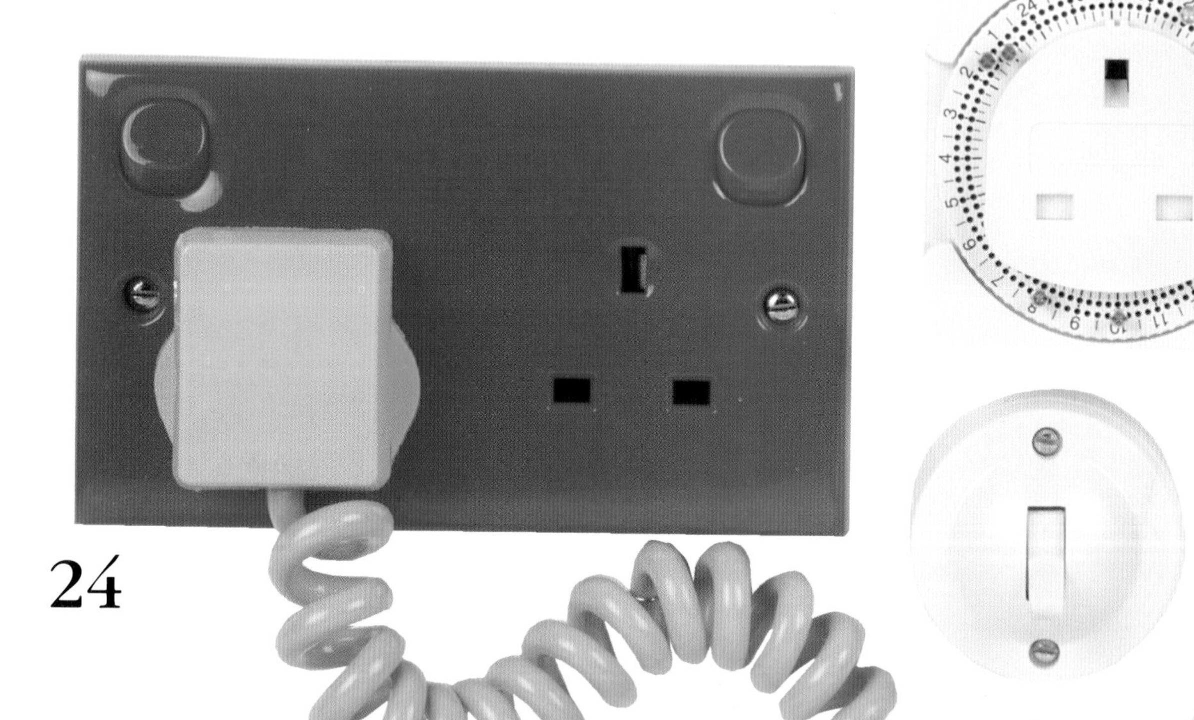

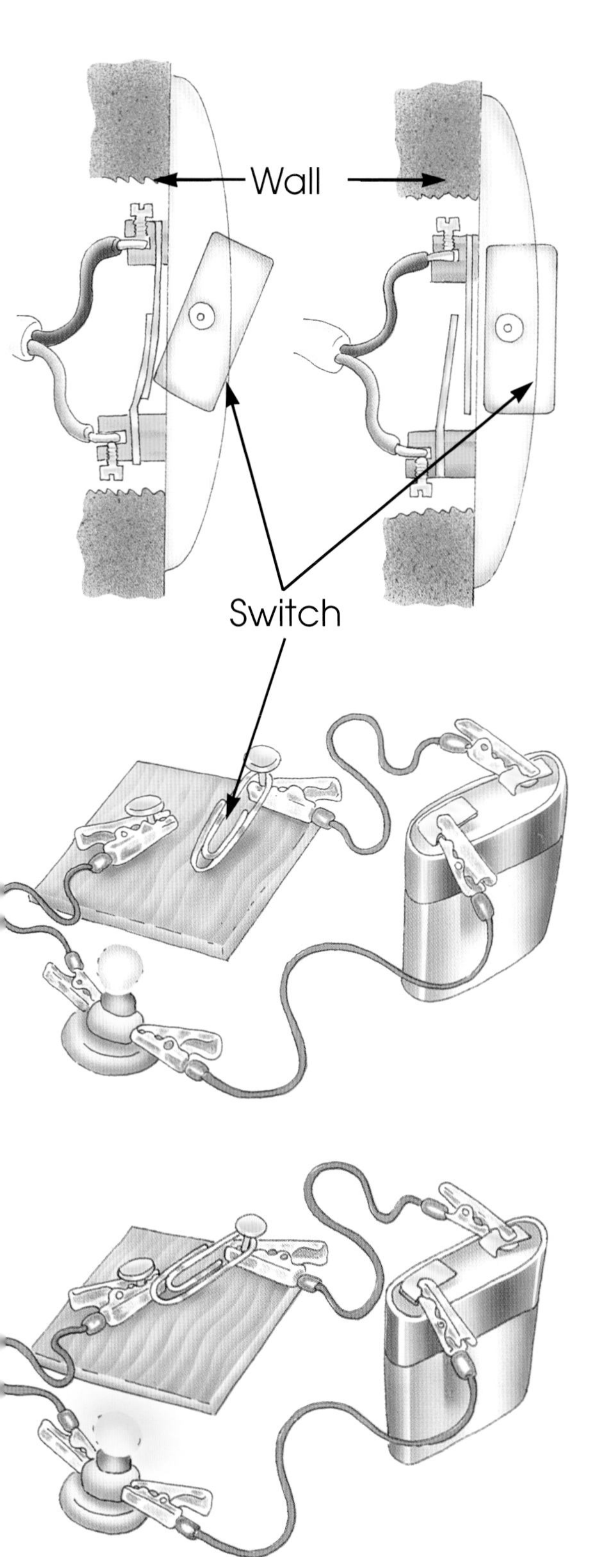

How do switches work?

Inside a switch there are two strips of metal. When you press a light switch off, the strips of metal cannot touch. This makes a gap in the circuit. The electricity stops flowing and the light goes out.

When you press the switch on, the strips of metal touch. They complete the circuit. Electricity flows and the light comes on.

Try adding a simple switch to your circuit. You could use a paperclip.

Light bulbs

Light bulbs come in many shapes and sizes.

How do bulbs work?

Look closely at a clear bulb, and you will see a thin wire running through it. Electricity flows along this wire when the light is switched on. The electricity makes the wire hot – so hot that it glows brightly and gives out light.

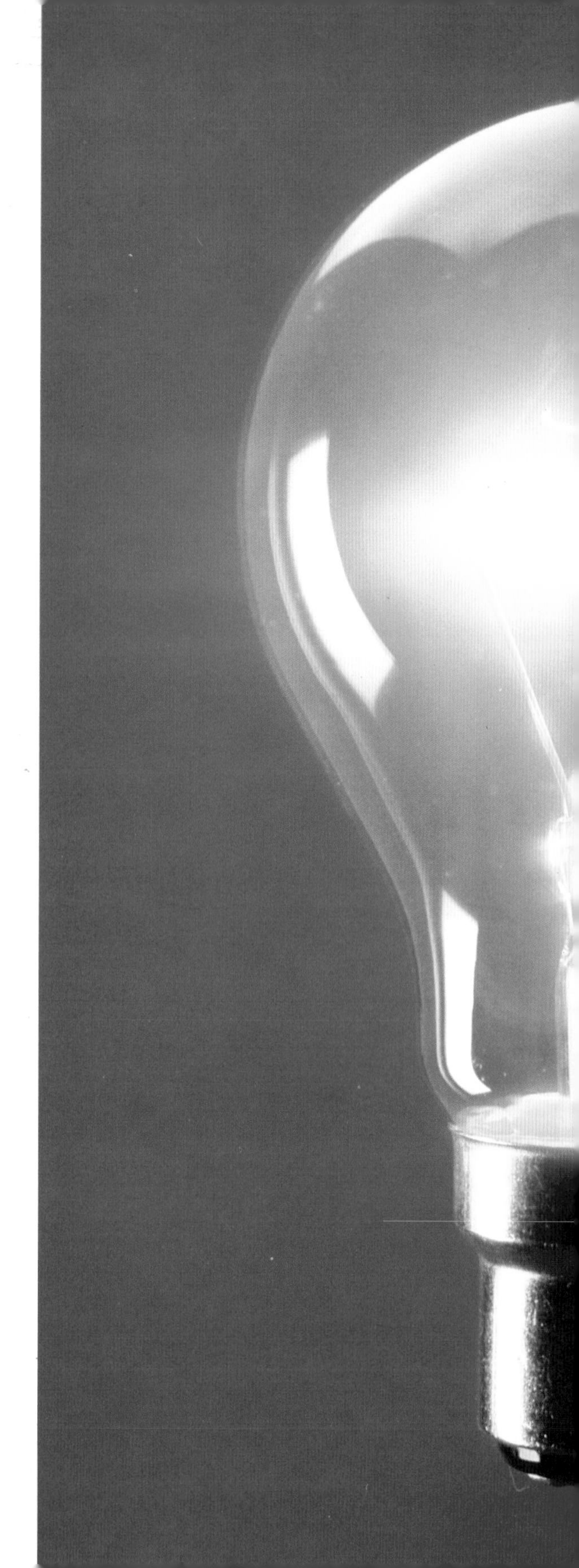

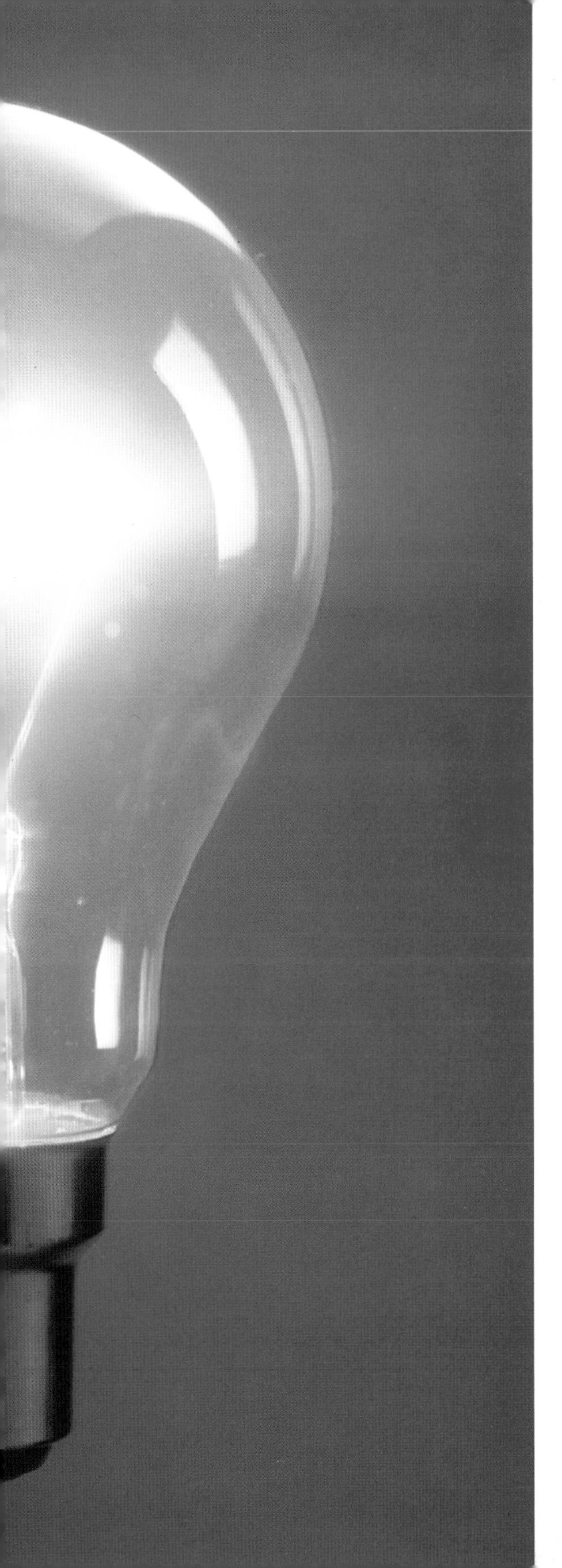

Two bulbs in a circuit

Try wiring two bulbs into a circuit using just one battery. You will need another piece of wire. What do you notice about the bulbs when they are lit?

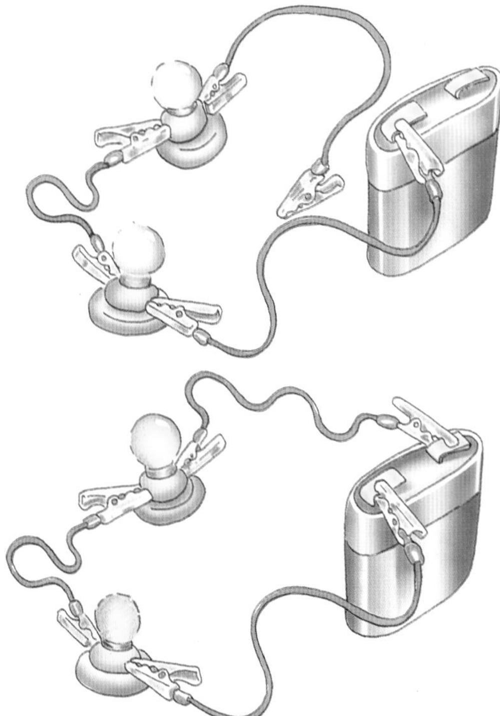

Paying for electricity

Electricity is not free. We have to pay for it.

How do we pay for electricity?
Every home has an electricity meter. The meter measures the amount of electricity that we use.
A person from the electricity company reads the meter, and sends us a bill.

What uses most electricity?
Some electric machines use much more electricity than others. Machines that make heat – like fires, cookers and kettles – use the most. Other machines – like a light or a radio – use only a little.

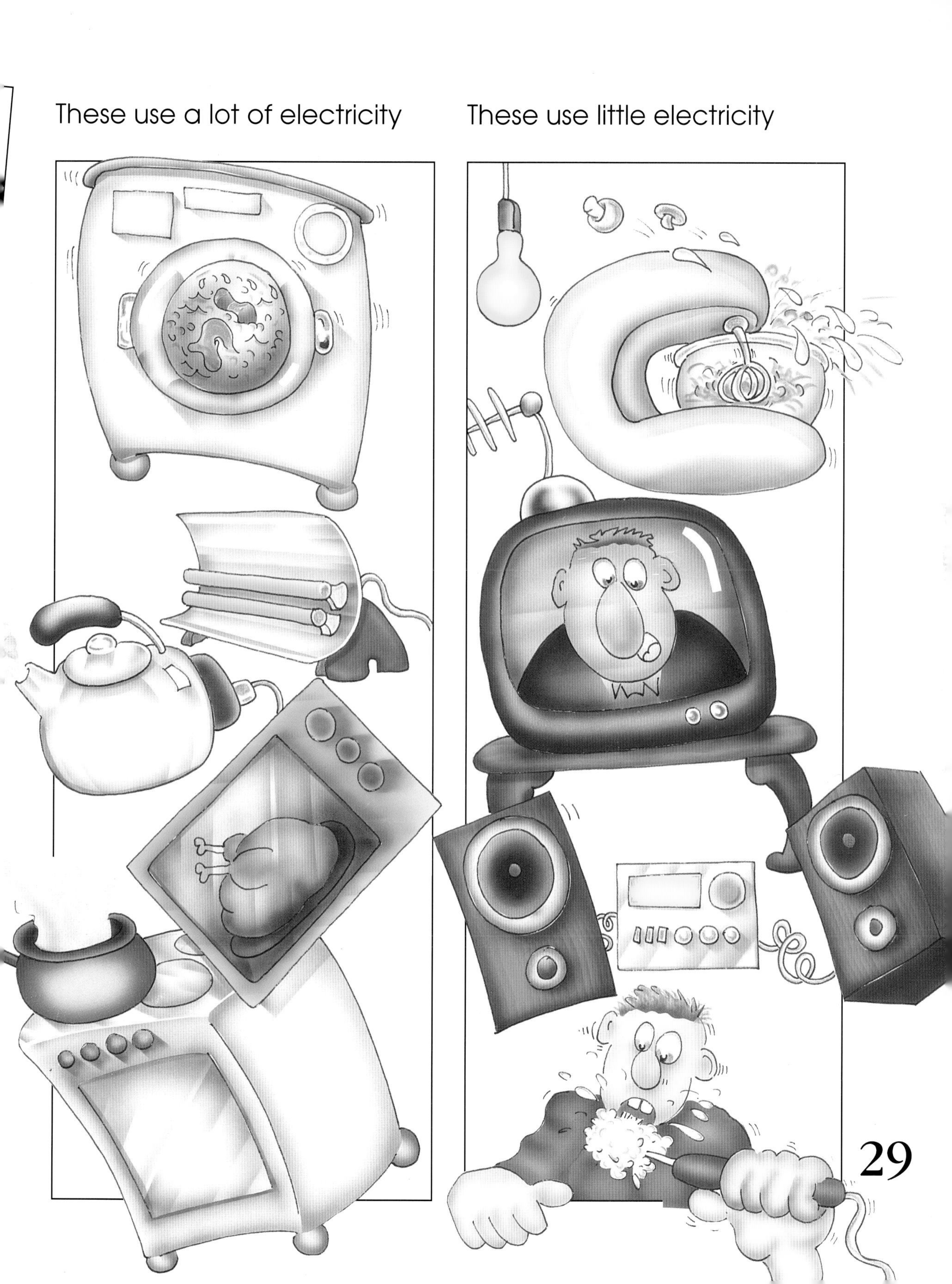
These use a lot of electricity
These use little electricity

Use it safely

Electricity is an important part of our lives. But it is very powerful, and can even kill. Always follow the safety rules.

Never play with wall sockets. The safe way to learn about electricity is with batteries.

Never use electric machines that have broken plugs or damaged flex.

Electricity and water are very dangerous together. Keep all electric machines away from water. Make sure your hands are dry before you touch anything electrical.

Never play with electricity. It can kill.

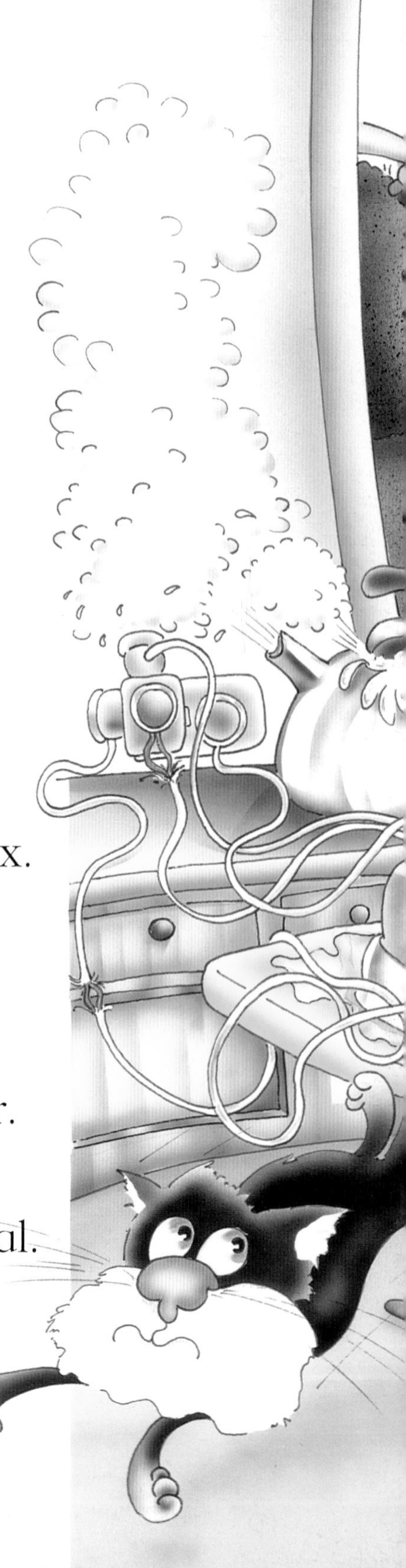

Index

HarperCollins Children's Books
A Division of HarperCollins Publishers Ltd, 77–85 Fulham Palace Road, Hammersmith, London W6 8JB
First published 1994 in the United Kingdom

Prepared by *specialist publishing services* 090 857 307

ISBN 0 00 196539 5
A CIP record is available from the British Library

Illustrated by Keith Hodgson
Photographs by Stephen Johnson/TSW: pp26/27; TSW: pp4/5,16; F Popley/Life File: p17; Mark Laing: pp12/13, 24
Series editor: Nick Hutchins; Editing: Claire Llewellyn; Design: Eric Drewery/Susi Martin; Picture research: Lorraine Sennett
Printed and bound in Hong Kong